D. Lothrop

Book of natural history

D. Lothrop

Book of natural history

ISBN/EAN: 9783741122750

Manufactured in Europe, USA, Canada, Australia, Japa

Cover: Foto ©berggeist007 / pixelio.de

Manufactured and distributed by brebook publishing software
(www.brebook.com)

D. Lothrop

Book of natural history

NATURAL HISTORY.

Boston:

Published by D. Lothrop & Co.

Dover, N. H.: G. T. Day & Co.

THE GUINEA-PIG.

I DON'T know why this animal should be called a pig, unless it be because it is covered with bristly hair. Its snout is not at all like that of a pig; it has feet like those of a rat, long whiskers, and short ears, but no tail. Besides this, it is not so big as a rabbit, and it does not grunt. Children are very fond of keeping Guinea-pigs in hutches, where they feed them on different kinds of vegetables, and often take them out to let them run about a

room, or carry them in their
arms.   They are gentle and
tame, but they cannot be taught
any pleasing tricks ; nor are
they of any use except to be
pretty playthings, with their
white hair neatly marked with
black and red.   In their native
country, which is South Amer-
ica, they are no doubt of some
use ; for GOD has made nothing
in vain. They live there in
the hedges of prickly pears, or
in the marshes where there are
plenty of juicy plants.   They
stay in their burrows all day,
and in the fine evenings come
out to feed ; and very pretty it

must be to see them moving quietly about, enjoying the warm weather and their supper of fresh salad.

THE SQUIRREL.

THE SQUIRREL.

NEAR the corner of my house stands a fine filbert-tree, which bears nearly every year a large number of nuts. One day in September, when my nuts were ripe, I happened to look out of the window, and saw in my tree a squirrel busily employed in gathering and eating the filberts. When I ·thought he had had enough, I opened the window and clapped my hands, upon which the squirrel skipped away, ran along the ground across my garden, and darting up the trunk of a tall fir-tree lay hid among the thick branches. Next day, I gathered all my nuts except a few on the top branches, which I left for the squirrel, or the dormouse, or the nuthatch, whichever should be the first to come. The squirrel had some, I know, for I found under the trees some halves of split shells. The dormouse had some, because some of the shells had round holes nibbled through them ; and the nuthatch had a share, because I saw the pretty bird busily at work cracking them. The squirrel and I had the most, because we laid some by in store against the winter.

THE WEASEL.

THE WEASEL.

THE weasel is a sly-looking animal, with a body well fitted for creeping into holes. It is a great enemy to rats, mice and moles, which it chases into their underground burrows. Though small, it is very fierce, and is not afraid of animals much larger than itself. These it kills by fixing its teeth in the back of their heads, and eating into them until they fall dead. Young hares and rabbits are often killed by it, and when it can get into a hen-roost, it sucks the eggs, and kills the young chickens and ducklings. It can climb trees and walls with ease, and when it finds a bird's nest, it kills the hen bird while sitting, and destroys the eggs or young birds. Some people say that it does more good than harm, by killing or driving away rats and mice, which are much afraid of it; but it is a sad character to be always hunting about for something to kill and eat. Another animal very like the weasel, only a little larger, is the stoat. This lives much in the same way as the weasel. In cold countries the skin of the stoat turns white in the winter, all except the tail, which is quite black. It is then called an ermine, and is much hunted for its fur, which is used in making the robes of kings and nobles.

THE TIGER.

THE TIGER.

WHAT a fierce and cunning-looking animal is this! We might fancy it to be a huge cat gone mad with disappointment and .rage. Think then how dreadful it must be to meet a wild beast like this. Thank God there are no such terrible animals in this country, though in India one must take good care how one ventures into the places where they are to be found. The tiger, though not swift, leaps with great strength, and his large head, heavy paws, and the great weight of his body enable him to spring on the head of the largest elephants, and fairly pull them down to the ground, riders and all. But the elephant sometimes shakes him off under his feet, and then either kneels on him and crushes him at once, or gives him a kick which sends him away limping. Some one on the elephant's back, most likely, has a gun ; then he tries to shoot him through the head, and many tigers have been killed in this way. The tiger is strong enough to kill and carry off to the woods a deer, a horse, or even a buffalo, and often does much mischief when it pays a visit to a village where cattle and flocks are kept; for it is too bold and savage to be driven away by men, and too cunning to be often caught in a trap.

THE KANGAROO.

THE KANGAROO.

" So you are called kangaroo, are you ? Your name. I think, ought to be little-head-and - great - tail. Your upper half seems as if it did not belong to the lower. And why do you stand in that awkward way, letting your fore-paws hang down like a pair of gloves ? And those heavy, clumsy hind legs ; what can you do with them ? How much better off you would be if your tail were cut off, and your legs shortened ! " " My friend," the kangaroo might say, if he could speak, "little people should look and learn. I do not hunt animals, so I have no need either of tusks or claws. When I graze I bend forward, and my mouth is close to the ground. If I should like to change my diet and eat a few leaves, I can sit up on my tail as long as I like. If I wish to go in search of new pasture I can amble along most pleasantly with my four legs, and I have no trouble in bringing my tail with me, heavy though you may think it. But when I am in a hurry, ah, you should see me then, and you would be surprised. By the help of my long legs and tail I can jump over bushes and streams, or over your head if you come in my way. It must be a very swift hound that can keep up with me.

THE SHREW.

. THE SHREW.

"Look here!" said little George to his uncle as they were walking together through a wood, "what is this curious little animal? It is very like a mouse, except that it is white beneath, and has a long, tapering snout. It is quite dead; I wonder what killed it!" "That," said uncle Willie, "is a shrew, a very useful little animal, because it eats a large number of insects which destroy the farmer's crops." "And can you tell me, uncle," said George, "how this one was killed?" "That I cannot well say. They are often found lying dead in the fields and lanes. The reason, perhaps, is, that the shrew has a very strong smell, like that of musk, which some animals of prey do not like; so, perhaps, this shrew was killed by a cat or stoat, but left uneaten. Cats will chase any small animal that they see moving, but they will not eat shrews, I know; for my cat often catches them, and brings them into the house; but she would rather have a fish's head for her dinner, no matter how stale it is, though that, in my mind, smells much worse."

THE ORANG-OUTANG.

THE ORANG-OUTANG.

This strange animal is called sometimes the wild man of the woods ; and, no wonder, for it has very much the look of a man. The country where it runs wild is very warm, so, when one of these animals was brought to England, it was clothed in a dress of flannel, that it might not suffer from the cold. This it liked very much, and did not try to tear it off as one might suppose it would do. It learned to drink out of a cup or glass, and to use a spoon for its food; and when it had done, it would carefully lay them down on the table or give them to its keeper. It was not given to be playful; it looks too grave for that perhaps you think; yet sometimes it would run after its keeper and pretend to fight with him; and if he went away, it would grow angry and try to break open its door. It was also very fond of playing with a cat, which it would carry about in its arms, whether pussy liked it or not, just as you, I dare say, sometimes carry about a

kitten. At night it would put
its bed to rights very carefully,
and cover itself with blankets
as snugly as you could.

THE HEDGEHOG.

THE HEDGEHOG.

HARRY MILLER was one eve-
ning walking in his father's
orchard, when he saw something creeping on the ground.
It was larger than a rat, but
not so big as a rabbit, and
seemed to have a very round
back. He stooped down to
feel it, but quickly drew back
his hand, for it pricked his
fingers. "Oh!" said he, "I
know what this is; I have
found a hedgehog. How glad
I am! I will take it home and
keep it in the kitchen, that it

may eat the crickets and black-beetles." But the hedgehog had rolled itself up into a prickly ball, and was not at all a pleasant thing to carry. Then he thought of a plan. He spread his pocket-handkerchief on the ground, and rolled the hedgehog into it with his foot. Then taking the four corners, he had his prize safe and ran merrily home. When he had reached home he placed it in a box for the night, and in the morning turned it out into a hole behind the fireplace in the kitchen. It soon grew tame, and lived on soaked bread,

potatoes and meat, besides the
insects which it caught for
itself.

THE REINDEER.

THE REINDEER.

A LONG way off in the north is a country called Lapland, a cold, dreary country, where there is no spring nor autumn, but only a long winter and a short summer. For months together the ground is white with snow, so that you would think that the people must all starve. But God takes care of His creatures as well in dreary Lapland as in happy England. There are great numbers of reindeer there, which are as useful to the Laplanders as cows, sheep and horses. Their flesh is as good as mutton at all seasons of the year. In summer they give plenty of milk, which their owners make into cheese for winter use, and when the ground is covered with snow, the Laplanders harness them to sledges — which are carriages without wheels—and travel as swiftly as horses in England draw a coach. But what do you think the reindeer do for food in the long, cold winter? Their masters have neither hay nor corn to give them, but God takes care of them. He has covered the ground with a plant like moss, which grows under the snow, and the reindeer, who are very fond of it, scrape off the snow with their strong hoofs, and find underneath all that they want.

THE MONKEY.

THE MONKEY.

Aн ! how I pity those poor
little monkeys we sometimes
see in the streets, dressed up in
dirty red clothes, and made to
dance or play tricks for the
amusement of idle people!
How frightened they look when
their master speaks to them
harshly, or when he lifts his
whip at them. I fear they are
not often treated kindly; in-
deed, it must be very cruel to
take them out at all into the
cold, wet streets, as we some-

times see them.   It is hard to
say whether they shiver from
cold or fright.   People must
either be ignorant or cruel to
take pleasure in such sports.
But to see monkeys skipping
about in the woods where they
were born is a very different
thing.   There they seem full
of fun and enjoyment, climbing
the trees as nimbly as squir-
rels, springing from branch to
branch, sometimes clasping
with their hands and sometimes clinging with their long
tails.   They live together in
large flocks, and when they
cannot find enough fruit in the

woods, they come by night into the corn-fields and gardens and do great mischief.

THE GIRAFFE.

THE GIRAFFE.

THIS animal is sometimes called the camel-leopard, because its head is like that of a camel, and its skin is spotted like a leopard. I once saw two tame giraffes in a garden where animals are kept. They were standing under a large ash-tree, from which they wanted to pluck some leaves; but, alas! there was not a leaf within reach; all the lower branches had been stripped long ago. What was to be done? One of them drew himself up to his full height, stretched out his long black tongue, and, having twisted it round a branch, pulled downwards.   The other giraffe was on the watch, and when the branch came near, caught hold of a leaf with his lips and pulled it off.   In doing this, he jerked the branch so that it uncurled the other giraffe's tongue and flew up again.   By the time that the leaf was eaten, the branch had been pulled down again, and another leaf was picked ; and so they went on for a long time, — one pulling down, the other eating.   But the one who

did all the work and got nothing for his pains was not at all
angry with the other. So the
giraffe must be a very kind and
gentle animal, though it is so
big. It is to be hoped that
they changed places after a
while; but this I could not stay
to see.

THE BEAVER.

THE BEAVER.

THERE are many animals which build nests for themselves and their young ones; but there is only one that builds itself a house with wood, stones and mud, and that is the beaver. Beavers are very fond of the water, so they make their houses, often many together, by the side of a river or lake. They cut down young trees with their sharp teeth, and carry the pieces in their mouths to the water's edge, where they lay them in order, and fill up the cracks with stones and mud, which they carry in their paws. It is in the summer that they do this. In autumn they lay up in their houses a store of bark, stripped from trees, for their food in winter. When cold weather comes, the mud freezes till it is almost as hard as stone; but the beavers do not mind, for they are safe within. They have no need to go abroad for food, and no wild animal can reach them, for the the door of their houses is under water. They can swim and dive famously, and with their broad, scaly tails they can

row themselves along as well
as a man can row a boat.
Their bodies are covered with
fine, soft fur, which is so very
useful that the poor beavers
have no peace when once the
hunters have found out where
they live.

THE RHINOCEROS.

THE RHINOCEROS.

" You have a very hard name, and I cannot call you pretty. You look like a great, awkward pig, wearing the skin of some other larger animal, and so it does not fit you well; but though pigs often have tusks in their under jaw, you carry yours on your nose." If the rhinoceros could speak, he would say : " Ugly though I am, I do no harm to any one if I am let alone. I live in a country where there are plenty of wild beasts who would soon

tear me to pieces, large as I
am, if my skin were not so
thick that they may scratch me
and bite me as much as they
please without hurting me. I
do not mind a spear, or an ar-
row, or even small shot from a
gun ; but why hunters try to
kill me I do not know, for I do
no harm to any one. My food
is reeds, leaves, and branches
of trees ; and I do not eat very
much of them considering my
size. When I cannot get any
other kind of food, I rip up the
trunks of trees, and, as my
teeth are very strong, I dine
off the green bark and wood.

I like nothing better than wal-
lowing in the mud on a hot
day ; but you had better keep
out of my way when I happen
to be angry, for, heavy though
I look, I can move pretty
quickly, and can use my horn
more ways than one.

THE MOUSE.

THE MOUSE.

THE mouse is a pretty little animal, but troublesome and mischievous. It is not choice in its food, but likes to taste everything, Flour, bread, candles, cheese, butter and twenty other things will surely be tasted if they come in its way. And it is of no use to wrap things in paper, or to put them away in wooden boxes, for mice, with their sharp teeth, soon nibble their way through. And they can climb so well, that it is hard to keep anything

out of their reach.   One night,
I was in bed, just going to
sleep, when I heard a mouse
creeping up the bed-curtains ;
I made a blow with my fist at
the place where the sound came
from, and drove it away.   But
it soon came back again, and
kept me awake so long that at
last I lighted a candle and
placed it on the chimney-piece,
thinking the light would drive
it away.   But soon I heard the
mouse creeping up the curtains
again, and then a sudden noise
in the candlestick.   The impu-
dent little fellow had crawled
to the top of the curtains and

sprung across to the chimney-piece ; and there it was, sitting up nibbling the drops of grease! This was too bad ; so I got up, opened the door, and hunted it out with a towel.